EARLY BIRD
ENERGY

HEAT

BY SALLY M WALKER
PHOTOGRAPHS BY ANDY KING

Lerner

LERNER BOOKS • LONDON • NEW YORK • MINNEAPOLIS

This book was first published in the United States of America in 2006.

First published in the United Kingdom in 2008 by
Lerner Books,
Dalton House,
60 Windsor Avenue,
London SW19 2RR

This edition was updated and edited for UK publication by Discovery Books Ltd., Unit 3, 37 Watling Street, Leintwardine, Shropshire SY7 0LW

British Library Cataloguing in Publication Data

Walker, Sally M.
  Heat. - (Early bird energy)
  1. Heat - Juvenile literature
  I. Title
  536

  ISBN-13: 978 1 58013 311 1

Additional photographs are reproduced with permission from: © Royalty-Free/CORBIS, pp. 5, 39; PhotoDisc/Getty Images, pp. 7, 13, 47. Illustrations on pp. 10, 14, 42 by Laura Westlund.

Printed in China

CONTENTS

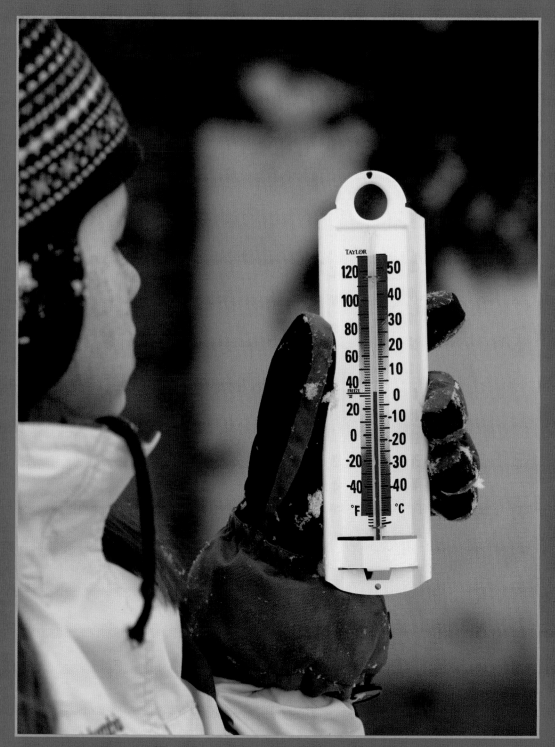

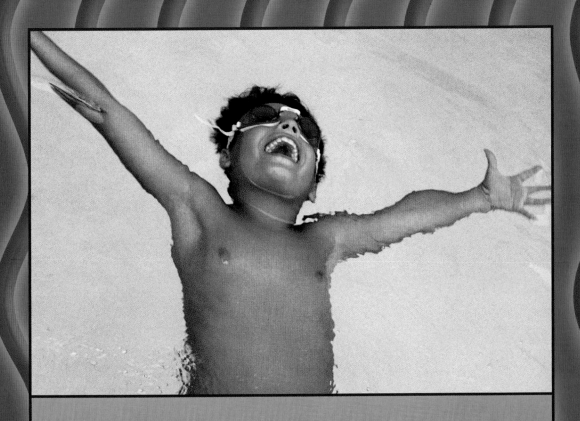

BE A WORD DETECTIVE

Can you find these words as you read about heat? Be a
detective and try to work out what they mean. You can
turn to the glossary on page 46 for help.

atoms	expands	melting
boiling	freeze	molecules
boiling point	freezing point	solids
condensing	gases	states
conducts	liquid	temperature
element	matter	thermometers
evaporation		

If a plant gets too hot or too cold, it dies. But the right amount of heat helps plants grow. What do we call the amount of heat an object has?

CHAPTER 1
WHAT IS HEAT?

Heat is important for all animals and plants. Without any heat, living things die. The same thing happens if living things get too much heat. But with the right amount of heat, animals and plants live and grow. Heat from the Sun helps keep living things healthy.

Heat is a form of energy. The amount of heat an object has is called its temperature. Water turns into ice at a temperature called the freezing point. Water turns into a gas at a temperature called the boiling point.

This desert is very hot in the daytime. Most plants and animals can't live there because it is too hot and too dry.

We measure temperature in units called degrees. Thermometers are tools that are used to measure temperature.

Some thermometers measure temperature with a scale called the Celsius scale. Using this scale, water freezes at about 0 degrees. Water boils at about 100 degrees. All scientists use the Celsius scale. So do most other people around the world.

In some places, the temperature stays below the freezing point of water for much of the winter. The water in lakes freezes into ice.

When water boils, it bubbles and steams.

Other thermometers measure temperature with a scale called the Fahrenheit scale. Using this scale, the temperature at which water freezes is about 32 degrees. Water boils at about 212 degrees.

TEMPERATURE SCALES

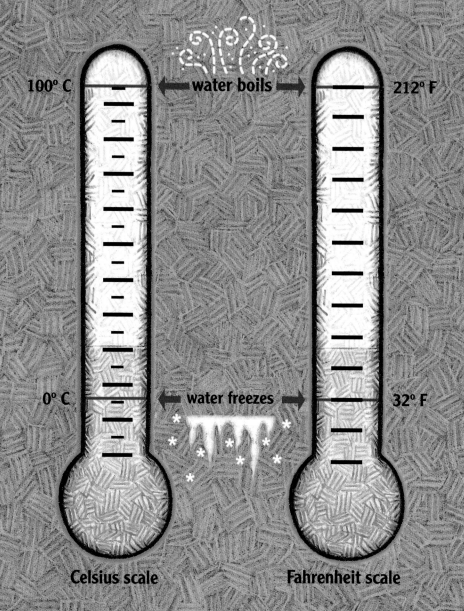

100° C ← water boils → 212° F

0° C ← water freezes → 32° F

Celsius scale   Fahrenheit scale

°C stands for Celsius degrees and °F stands for Fahrenheit degrees.

There are many different kinds of thermometers. Some measure the temperature of a room. Some are used to find out if a sick person has a fever. Some are used to measure the temperature of very hot things, such as ovens or melted rocks. Some measure the temperature of cold places, such as freezers.

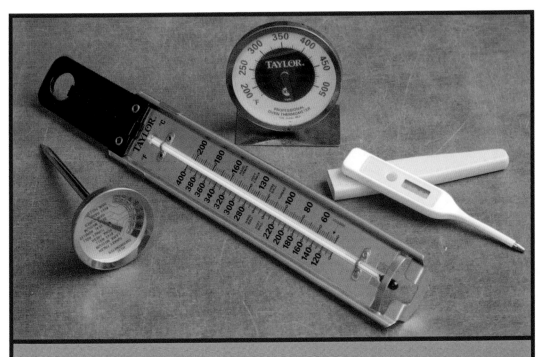

The thermometer on the left is used for cooking meat. The long metal thermometer is used in making sweets. The thermometer at the top measures an oven's temperature. The white plastic thermometer is used to take a person's temperature.

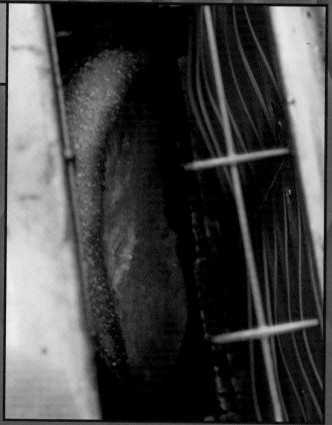

Toasters use heat to make bread brown and crunchy. What are some other ways that we use heat?

CHAPTER 2
WHAT MAKES HEAT?

Boilers and electric heaters heat our houses. Stoves and ovens heat our food. Engines in cars make heat when they run. But where does heat come from? The answer begins with matter.

All matter takes up space and can be weighed.

Matter is anything that takes up space and can
be weighed. All of the objects around you are
made of matter. Books, pencils, air and milk
are all matter.

Matter is made up of very tiny particles called atoms. Atoms are so small that billions of them can fit on the full stop at the end of this sentence.

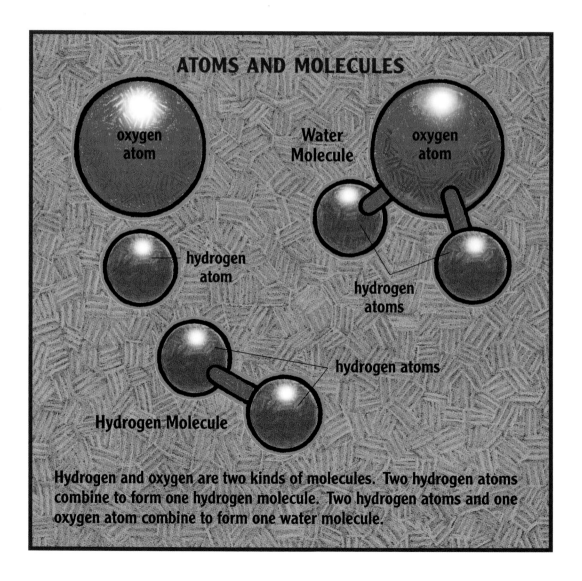

ATOMS AND MOLECULES

oxygen atom

Water Molecule

oxygen atom

hydrogen atom

hydrogen atoms

hydrogen atoms

Hydrogen Molecule

Hydrogen and oxygen are two kinds of molecules. Two hydrogen atoms combine to form one hydrogen molecule. Two hydrogen atoms and one oxygen atom combine to form one water molecule.

The dark centre part of a pencil is called the lead. Pencil leads are made of an element called carbon.

There are more than 100 different kinds of atoms on Earth. Each kind of atom is called an element. Oxygen, gold, helium, hydrogen and calcium are some of the elements. Atoms can join together to form groups called molecules. Some molecules are made up of only one kind of atom. The molecules in pure hydrogen gas are all made of hydrogen atoms.

Sometimes water drips or flows. But even when water is still, its molecules are moving.

But most molecules form when different kinds of atoms join together. This makes a new substance. When two atoms of hydrogen join with one atom of oxygen they make one molecule of water.

Atoms and molecules are always moving. When they move, they make heat. You can't see heat, but you can feel it. You can see how it changes the objects around you.

Hot molecules move faster than cold molecules. You can prove it with a simple experiment. You will need two bowls and some food colouring.

Fill one bowl with ice-cold water. Fill the other bowl with very hot water from the tap. Be careful that you don't scald yourself.

Let the water run for a bit before you fill each bowl. It may take a minute or two for the water to become very cold or very hot.

Add one drop of food colouring to each bowl. Do not stir the water. What happens? Does the food colouring spread out faster in the hot water or the cold water? It spreads faster in the hot water.

Add a drop of food colouring to each bowl.

The food colouring spreads out more quickly in hot water *(right)* than in cold water *(left)*.

Wait two minutes. Look in the bowls again. The heated water is almost completely coloured. The hot molecules are moving fast. So the molecules of food colouring spread out quickly. But not all of the icy water is coloured. Its cold molecules are moving slowly. It takes much longer for the molecules of food colouring to spread out.

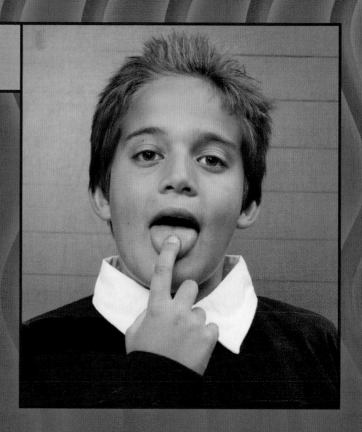

CHAPTER 3

CHANGING FROM HOT TO COLD

Matter can be hot, cold or somewhere in between. Touch your tongue. Your tongue is warm matter. Molecules moving inside your body make a lot of heat. That's why your tongue is warm.

Ice cubes are cold. Their molecules don't move much.

An ice cube is colder than your tongue. The molecules in ice barely move. They make almost no heat.

Heat always moves from warmer matter to colder matter. When you wrap your hands around a mug of hot chocolate, heat leaves the hot chocolate. It moves into your hands. The hot chocolate becomes cooler, and your hands become warmer.

Drinking hot chocolate helps you warm up on a cold day.

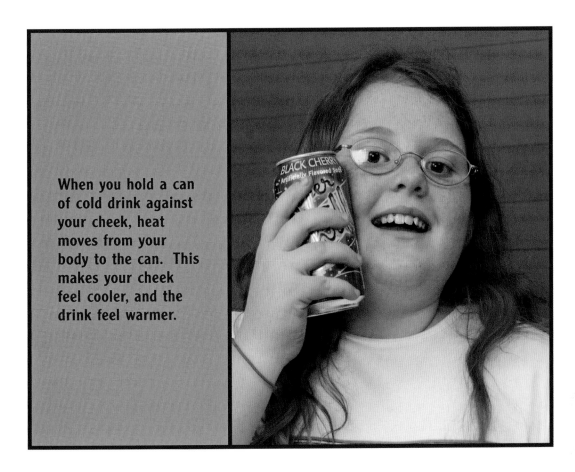

When you hold a can of cold drink against your cheek, heat moves from your body to the can. This makes your cheek feel cooler, and the drink feel warmer.

The opposite happens with a can of cold drink on a hot summer's day. When you put the cold can against your hot cheek, heat flows away from your cheek. You feel cooler because heat is leaving your cheek. Heat moves from your cheek to the can of drink. The drink and its can become warmer.

A popped kernel of popcorn is much bigger than an unpopped kernel. What makes the kernel get bigger?

CHAPTER 4

HEAT MAKES MATTER GET BIGGER

When matter is heated, it expands. When matter expands, it gets bigger. Gas expands when it is heated. You can prove this. You will need a drinking straw and a tall glass.

Be sure to keep the end of the straw tightly covered for the whole experiment. If your finger slips, you will need to start again.

Fill the glass with very hot water from the tap. Be careful not to scald yourself. Hold your finger tightly over one end of the straw. This stops the air inside the straw from escaping from the hole at that end.

Keep your finger over the hole. Dip the other end of the straw into the water. Watch the end of the straw that is in the hot water. Slowly push the end of the straw almost to the bottom of the glass. What happens?

Push the straw down until it almost touches the bottom of the glass.

If you don't see a bubble of air, try again. Make sure the water in the glass is very hot. And don't let your finger slip off the end of the straw.

An air bubble forms at the uncovered end of the straw. When you put the straw in the water, heat from the water warms the air inside the straw. The air starts to expand. It needs more space. But it can't push out of the top of the straw. Your finger is in the way. Instead, the heated air pushes downwards into the water.

Solid matter also expands when it's heated. Some pavements are made of slabs of concrete. You may have noticed that there are spaces between the slabs. In the summer, the Sun heats the concrete. The heat makes the pavement expand. The spaces give each pavement slab room to expand. If builders did not leave the spaces, the concrete would crack.

The spaces between pavement slabs give the slabs room to expand.

Builders leave spaces between the parts of a bridge.

A bridge made of metal and concrete also has spaces between its parts. The spaces give the parts of the bridge room to expand as the seasons change. Without them, the bridge would crack apart.

Liquid matter expands when it is heated too. You can prove it with an outdoor thermometer and a bowl of warm water. Look at the thermometer's glass tube. At the bottom is a red or silver bulb. Can you see a thin red or silver line inside the tube? It comes up from the red or silver bulb. The colour you see is liquid that is sealed inside the tube. How far up the tube does the coloured line go?

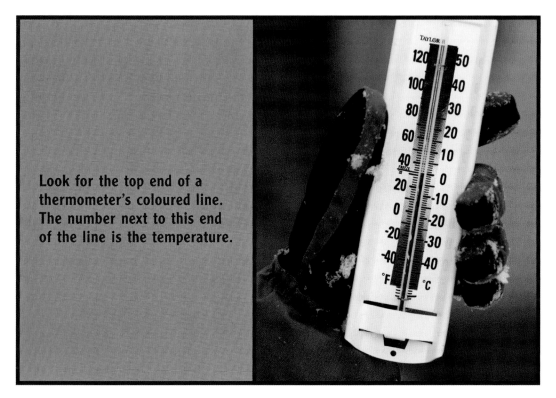

Look for the top end of a thermometer's coloured line. The number next to this end of the line is the temperature.

Some thermometers have a needle instead of coloured liquid. The needle moves to point at the temperature.

Put the bulb of the thermometer in the bowl of warm water. Watch the coloured line. What does it do? The line gets longer. Why? Because the liquid matter inside the tube is expanding. As it expands, the liquid rises higher inside the glass tube.

If you need to pick up a hot pan, wear oven gloves or use pot holders. Oven gloves and pot holders protect your skin so you won't be burned.

CHAPTER 5
HOW HEAT MOVES

Matter conducts heat. This means that matter lets heat move through it. Some kinds of matter conduct heat faster than others. You can prove this yourself. You'll need a tall glass, scissors, a piece of cloth, aluminium foil, some sticky tape and a foam food tray.

Tape the foil, foam and cloth strips around the glass.

Cut a strip from the foil. The strip should be 2.5 centimetres wide and long enough to wrap around the glass. Then cut a strip from the cloth and one from the foam tray. These strips should be the same size as the foil strip. Tape the three strips around the glass.

Fill the glass with very hot water from a tap. Be careful you do not scald yourself. Slowly count to 15. Then feel each strip. Which strip is hottest?

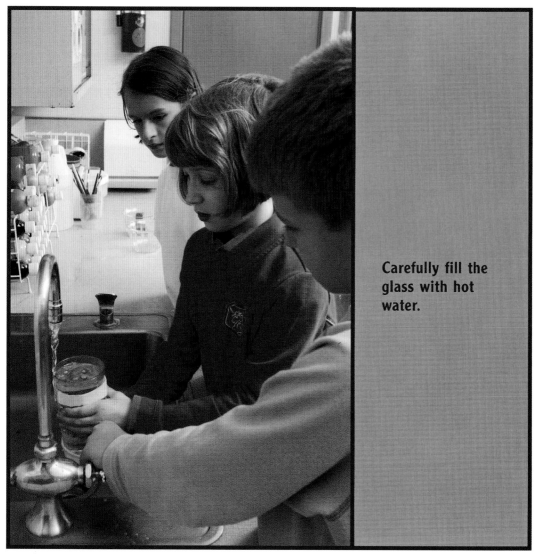

Carefully fill the glass with hot water.

Touch each of the three strips. Which one feels hottest? Which one feels coolest?

The foil is hottest. Its molecules conduct heat quickly. The cloth strip is warm. Its molecules conduct heat more slowly than the foil does. The foam is coolest. Foam conducts heat very slowly. The reason is that foam has lots of air bubbles inside it. Air conducts heat much more slowly than foil or cloth does.

Foil is made of a metal called aluminium. Metals are good heat conductors. That is why we use metal pans for cooking. Heat from a stove moves quickly through a metal pan. So the food in a metal pan heats up quickly. That's a good thing when you are hungry.

To cook food, we must add heat to it. Food heats up quickly in a metal pan. So the food cooks quickly too.

Heat moves from warm air into cold drinks. The heat makes the drinks warmer. A foam cooler doesn't conduct heat from the air. So a foam cooler keeps drinks cold.

Foam isn't a good heat conductor. Foam is often used in the walls of houses. The air bubbles in the foam keep heat from moving quickly through it. This helps to keep heated air inside a house. So you stay warmer in cold weather!

Wood is solid matter. What kind of matter is milk?

CHAPTER 6
HEAT CHANGES MATTER

Matter comes in three forms called states. The three states are solids, liquids and gases. Wood is a solid. Milk is a liquid. The air we breathe is a gas.

Ice melts when heat is added to it. The solid ice becomes liquid water.

Adding heat can make matter change from one state to another. Adding heat to a solid turns the solid into a liquid. This is called melting. Adding heat to a liquid turns the liquid into a gas. This is called boiling.

The inside of a volcano is so hot it can melt solid rock into liquid. This liquid is called magma.

Adding lots of heat to water makes it boil. The water changes from a liquid into a gas called water vapour. This is called evaporation.

When water becomes very hot, it boils. The liquid water becomes a gas.

Air contains water vapour. Heat moves from warm water vapour into a cold glass. When the water vapour loses heat, it changes into drops of liquid water.

Losing heat can also make matter change its state. When a gas loses heat it becomes a liquid. As water vapour loses heat it becomes a liquid. Water is liquid matter. When a liquid loses heat, it becomes a solid. If you remove enough heat from water, it freezes. It becomes solid matter called ice.

CHANGES OF STATE

steam
(gas)

Adding heat to a liquid
turns it into a gas.
This is called
boiling.

Taking heat away
from a gas turns the gas
into a liquid. This is called
condensing.

water
(liquid)

Taking heat away from a
liquid turns the liquid into a
solid. This is called
freezing.

ice
(solid)

Adding heat to
a solid turns it into
a liquid. This is called
melting.

Heat from the Sun makes a cold day feel warmer.

You have learned a lot about heat. Although you can't see heat, you know it is there. It is inside you and everything you see or touch. Explore your home, school and neighbourhood. Be a heat detective. Look for heat clues all around you.

A NOTE TO ADULTS
ON SHARING A BOOK

When you share a book with a child, you show that reading is important. To get the most out of the experience, read in a comfortable, quiet place. Turn off the television and limit other distractions, such as telephone calls. Be prepared to start slowly. Take turns reading parts of this book. Stop occasionally and discuss what you're reading. Talk about the photographs. If the child begins to lose interest, stop reading. When you pick up the book again, re-read the parts you have already read.

BE A VOCABULARY DETECTIVE

The word list on page 5 contains words that are important in understanding the topic of this book. Be word detectives and search for the words as you read the book together. Talk about what the words mean and how they are used in the sentence. Do any of these words have more than one meaning? You will find the words defined in a glossary on page 46.

WHAT ABOUT QUESTIONS?

Use questions to make sure the child understands the information in this book. Here are some suggestions:

> What did this paragraph tell us? What does this picture show? What do you think we'll learn about next? What happens when matter is heated? Why are some pavements made with spaces between the slabs? Why does heat move slowly through foam? What are the three states of matter? What is your favourite part of the book? Why?

If the child has questions, don't hesitate to respond with questions of your own, such as: What do *you* think? Why? What is it that you don't know? If the child can't remember certain facts, turn to the index.

INTRODUCING THE INDEX

The index helps readers find information without searching through the whole book. Turn to the index on page 48. Choose an entry such as *freezing* and ask the child to use the index to find out at what temperature water freezes. Repeat with as many entries as you like. Ask the child to point out the differences between an index and a glossary. (The index helps readers find information, while the glossary tells readers what words mean.)

LEARN MORE ABOUT HEAT

BOOKS
Hunter, Rebecca. *Hot and Cold* (Discovering Science) Raintree
    Publishers, 2003.

Royston, Angela. *Hot and Cold* (My World of Science) Heineman
    Library, 2004.

WEBSITES
Go Figure—How Popcorn Pops
http://www.kidzworld.com/site/p547.htm
Find out how heat makes popcorn pop.

Heat
http://www.chemistry.org/portal/a/c/s/1/wondernetdisplay.html?DOC
=wondernet\activities\heat\heat.html
This site has several activities that show how heat works.

Materials: Solids, Liquids and Gases
http://www.bbc.co.uk/schools/revisewise/science/materials/08_act.shtml
This site has information about the states of matter, plus activities
    and quizzes.

Thermometer
http://pbskids.org/zoom/activities/sci/thermometer.html
Learn how to make your own thermometer.

GLOSSARY

atoms: the tiny particles that make up all things

boiling: changing from a liquid into a gas

boiling point: the temperature at which water turns into a gas

condensing: changing from a gas into a liquid

conducts: lets something move through. Matter conducts heat.

element: a substance that cannot be broken down into different substances because it is made of only one kind of atom

evaporation: when a liquid is heated and changes into a gas.

expands: becomes bigger. When matter is heated, it expands.

freeze: to change from a liquid into a solid

freezing point: the temperature at which water turns into ice

gases: substances that can change their size and their shape. Air is a gas.

liquid: a substance that flows easily. Water is a liquid.

matter: what all things are made of. Matter takes up space and can be weighed.

melting: changing from a solid into a liquid

molecules: the smallest pieces of a substance. A molecule is made up of atoms that are joined together.

solids: substances that stay the same size and the same shape. Wood is a solid.

states: the solid, liquid and gas forms of matter

temperature: the amount of heat an object has

thermometers: tools that are used to measure temperature

INDEX

Pages listed in **bold** type refer to photographs.